IMAGES
of America

BRIDGES OF THE MID-HUDSON VALLEY

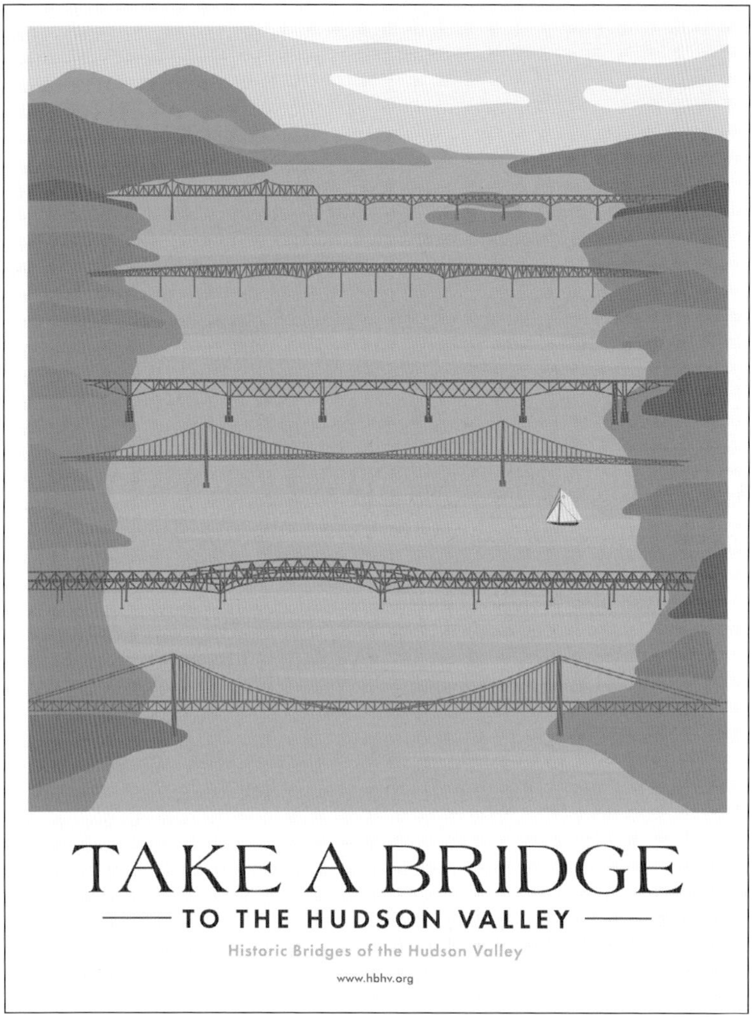

Historic Bridges of the Hudson Valley is a not-for-profit established in 2013 to educate on the unique and historic structures of the bridges of the New York State Bridge Authority, displayed beautifully in this image. Historic Bridges of the Hudson Valley provides historical programs, STEM and STEAM education and curriculum for area educators and students, access to our bridges for education, and much more. (Historic Bridges of the Hudson Valley.)

On the Cover: The city of Poughkeepsie in the 1950s was a bustling place. A woman stands in the roadway attempting to cross the traffic lined up to cross the Mid-Hudson Bridge, seen in the distance. A widened approach on the Poughkeepsie side will soon be configured to allow two to three lanes of traffic. Further traffic studies will result in the development of an elevated approach to the bridge, enabling free movement of traffic on the busy Route 9. (New York State Bridge Authority.)

IMAGES
of America

BRIDGES OF THE MID-HUDSON VALLEY

Kathryn W. Burke
Foreword by Tara Sullivan

ARCADIA
PUBLISHING

ISBN 978-1-4671-0542-2

Published by Arcadia Publishing
Charleston, South Carolina

Printed in the United States of America

Library of Congress Control Number: 2020932388

For all general information, please contact Arcadia Publishing:
Telephone 843-853-2070
Fax 843-853-0044
E-mail sales@arcadiapublishing.com
For customer service and orders:
Toll-Free 1-888-313-2665

Visit us on the Internet at www.arcadiapublishing.com

To all the members of the New York State Bridge Authority, past and present; to my husband Walter, daughters Maureen and Kate, and to Chris, Nick, Maeve, Walter, and the twins

Contents

FOREWORD

From "Rip" to "Bear," the five iconic Mid-Hudson Valley Bridges are beloved by every friend, family member, and neighbor I have in the valley. More interesting is how proud and proprietary the regional people and travelers are of "their" bridges.

They, like me and like the author of this book, see these bridges as integral to their environment as the Catskill Mountains, the Hudson Highlands, and the mighty Hudson River. Their bridges are part of their history, of their parents and grandparents telling stories of when they were designed by the most famous engineers of that time, that the bridges "were opened by Rockefellers and Roosevelts," and how they opened the movement of American travel and commerce westward. Their bridges are the economic lynchpin of the entire valley, with the lowest tolls in the nation so as not to create a barrier to traveling east to west and with a consistent toll among all of them to not tempt travelers to move north or south to the cheaper bridge.

And all of this ingenious and perfect function is due to the little regional authority that runs them with 196 all-in and dedicated employees—some who have stayed on for 30, 40, and even 50 years—demonstrating the true meaning of public service.

The New York State Bridge Authority, founded in 1932, has owned and operated these gems of infrastructure for nearly 90 years under an enviable motto, "Maintenance deferred is maintenance denied." Enviable? If all of America's bridges and roads were maintained on this principle, we would not have the constant headlines decrying the crumbling infrastructure in our country.

More amazing is this maintenance principle has merited the highest structural safety rating for the best-maintained bridges in the nation. More amazing still, the bridge authority does this with one of the lowest tolls in the nation. When do we ever see the very best for the lowest price?

Kathy Burke, the author of this book as well as *Hudson River Bridges*, is the oracle and unique connection among these bridges, the regional authority that has owned and maintained them since 1932, and the community that uses and "owns" them.

As a founding member and the first director of the nonprofit Historic Bridges of the Hudson Valley, she envisioned the lasting impact of raising the consciousness of students, parents, and civic leaders beyond the simple value of a road surface transporting them over the cold waters of the Hudson. With the establishment of a "teaching" bridge museum and classrooms on the bridges, historical interpretive signs on all of the bridges, and annual commemorative events on the bridges, she, like the bridges, is a great "connector."

As the first female leader of the bridge authority, I have had the great honor of working with Kathy and the dedicated staff of the authority to keep this unique and sterling legacy alive and well in the Hudson Valley.

May it never change.

—Tara Sullivan

ACKNOWLEDGMENTS

The following people deserve many thanks for their assistance in making this book possible: from the New York State Bridge Authority (NYSBA), Tara Sullivan, acting executive director and first female to lead that organization, for her continued support and for writing the foreword for this book; NYSBA engineers George Fong and Joseph Rochfort and chief engineer Jeff Wright, for their technical assistance, and to Jeff Wright for taking the time to review all proofs for engineering accuracy; to Greg Herd for his amazing photograph of the Mid-Hudson Bridge necklace lights and to all authority members who provided photographs and allowed themselves to be included in this book; and lastly, to Jade Mitchell, much more than an intern, for her help finding details of the history and locating things I couldn't find—she is much appreciated. Thank you also to Barney Martin, of Modjeski & Masters, for providing the construction photographs of the Mid-Hudson Bridge from the Modjeski & Masters archive; to Kay Peterson at the National Museum of American History (NMAH) at the Smithsonian, for all her assistance with archived photographs; to Geoff Brault and Fred Schaeffer at the Walkway Over the Hudson, for the photographs and content for the book; to Joseph Bertolozzi, for his contributions to text and photograph of *Bridge Music*; to Sarah Nivens at the FDR Presidential Library, for the photographs of the opening ceremony of the Mid-Hudson Bridge; to Robert and Loretta Gould, for allowing the inclusion of the painting of Robert's father, artist John Gould; and to anyone else I might have forgotten to mention, thank you. I'm eternally grateful for the loving support of my family: Walter, my husband; and Maureen, Chris, Kate, Nick, Maeve, Walter, and the twins. And lastly, at Arcadia Publishing, thank you to Erin Vosgien, who was my editor on my first *Hudson River Bridges* book and the person I reached out to when considering writing this second book, and finally, to Caroline Anderson, my editor for this book, who was always available to answer questions and provide assistance and support. I would also note that all photographs in this book, other than the Modjeski & Masters archive photographs of the Mid-Hudson Bridge and any photographs not otherwise credited, were obtained from the archives and files of the New York State Bridge Authority.

INTRODUCTION

Living in the Hudson Valley of New York State, it's easy to recognize the importance to the valley of the five crossings of the New York State Bridge Authority, but it is not as easy to recognize the importance of the New York State Bridge Authority to the bridges. The bridge authority is a small agency relative to other transportation agencies. It goes about its day-to-day business maintaining and planning for the long life of seven bridge structures, one of which is three decades over 100 and two of which are very nearly 100 years old.

The story of these Hudson River bridges is the story of the New York State Bridge Authority. It begins with the first bridge to come under the care of the bridge authority, the Mid-Hudson Bridge, and ends with the last bridge acquired, the historic structure of the Walkway Over the Hudson, the old Poughkeepsie railroad bridge. The bridge authority was created in 1932 by New York State government during the Great Depression, when state funds were not available to build the Rip Van Winkle Bridge. Then Gov. Franklin D. Roosevelt vetoed the bill to build the bridge, thus making the way to create the bridge authority.

The Mid-Hudson Bridge was built out of grass-roots community efforts. A local priest, Monsignor Joseph F. Sheahan, led the movement tirelessly until the Hudson Valley Bridge Association was formed, and the rest is history. It was a greater community effort, bringing together people from both sides of the river—the city of Poughkeepsie on the east and the farmlands on the west. The produce industry west of the river at that time, according to a report by the Hudson Valley Bridge Association, was shipping between $15 and 20 million worth of fruit and vegetables yearly. It was more productive than any region of the United States except Southern California. Add to that the Bronx Terminal Market, with funding provided in 1922 and projected to open in the early 1930s. The newest transportation available for growers, the motor truck, currently had to drive north to Albany to cross the Hudson River. Automobiles were becoming more affordable and therefore more popular. As with all the river crossings to come, the grassroots efforts of local communities were responsible for getting a bridge built across their section of the Hudson.

The Mid-Hudson Bridge was designed by Ralph Modjeski with Daniel E. Moran, both eminent engineers of the time. Modjeski partnered with Moran, as Moran had been an engineer for the substructure of the Poughkeepsie railroad bridge. From that experience, it was known that the bottom of the Hudson River was a difficult place to construct bridge foundations, and Modjeski wanted a partner with experience with the river bottom in that area of the Hudson. The New York State Department of Public Works, headed by Superintendent Frederick Stuart Greene, would be in charge of the building of the bridge. The Department of Public Works would later become the New York State Department of Transportation, which oversees all transportation-related construction in New York State. The bridge was begun during the administration of Gov. Al Smith, who laid the cornerstone in 1925. It was completed five years later, in 1930, during the administration of Gov. Franklin D. Roosevelt. The Mid-Hudson Bridge has become an iconic symbol of the Mid-Hudson Valley.

The Rip Van Winkle Bridge, north of the Mid-Hudson Bridge, crosses the Hudson River between the village of Catskill on the west and just south of the city of Hudson on the east. Grassroots efforts brought about the construction of this bridge, too. The hope was to build a bridge to connect not just the communities directly on each side of the river but also to connect this area of the Hudson Valley to New England by way of new routes; unfortunately, that hope was never actually realized. However, built in the "Birthplace of American Art," the beautiful structure of the Rip Van Winkle Bridge would add to a bucolic setting that had inspired the likes of Thomas Cole and Frederic Church of the Hudson River School, the first American art movement. The Rip Van Winkle Bridge, opened in 1935, quickly became known as one of the most scenic crossings in America. In 2019, the Hudson River Skywalk opened, much of which crosses the Rip Van Winkle Bridge. The bridge authority constructed three viewing decks along the sidewalk of the bridge. The viewing decks are each 82 feet long and extend four feet out from the sidewalk, providing spectacular views of the Hudson River, the Hudson Valley, and the mountains beyond.

Acquired by the bridge authority in 1940, the Bear Mountain Bridge, the southernmost bridge of the bridge authority, was built by the Bear Mountain Hudson River Bridge Company. The Harriman family had created the popular Bear Mountain Park in the Hudson Highlands and desired to build a bridge over the river to provide better access to the park. Roebling Sons and Company, builder of the Brooklyn Bridge, was hired to provide the steel cable for the bridge. The construction of the Bear Mountain Bridge became an engineering study in the new method of spinning parallel wire cables to create the main suspension cables of a suspension bridge. When it opened in 1924, it was the longest suspension bridge in the world and the first bridge across the Hudson River south of Albany.

The Kingston-Rhinecliff Bridge, designed by D.B. Steinman before he went on to design the famed Mackinac Bridge over the Mackinac Strait in Michigan, was built at the time of the development of the interstate highway system. It was designed with a sleeker, modern look. The cantilevered, continuous underdeck truss bridge afforded spectacular views and access to the Catskill Mountains to the west and the Berkshire Mountains to the east. The Kingston-Rhinecliff Bridge opened in 1957 and connects Kingston with the town of Rhinebeck.

The twin spans of the Newburgh-Beacon Bridge, the newest structures of the bridge authority, carry Interstate 84, connecting New England with points west. The north span, opened in 1963, soon became crowded with traffic as Interstate 84 was completed. The construction of the south span, opened in 1980, would help with the congestion, but the widening of the north span, reopened in 1984, would complete the crossing. The north span carries traffic westbound on Interstate 84, and the south span carries traffic eastbound. Both spans were designed and construction was overseen by the engineering firm of Modjeski and Masters, the firm created by Mid-Hudson Bridge engineer Ralph Modjeski.

The last bridge acquired by the bridge authority was the structure of the Walkway Over the Hudson in 2010. Known for its excellent maintenance record on its bridges, the State of New York transferred the responsibility for maintaining the 130-plus-year-old structure to the New York State Bridge Authority. The bridge structure of the Walkway immediately became a part of the 5- and 20-year maintenance plans of the authority, providing the necessary maintenance to keep the more-than-a-century-old structure healthy.

The popularity of the automobile necessitated the building of the bridges. The Hudson River, which had been a main transportation thoroughfare for so many years for freight and people, had become a barrier, and it must be crossed. Ferries, in some locations in use for hundreds of years, were limited to seasonal use and could not run when the ice formed on the river. Long lines of cars waiting to cross gave evidence of the inefficiency of ferry service. Each community in its time was provided its crossing. The bridges in existence now, from the Bear Mountain in the south to the Rip Van Winkle in the north, with careful maintenance of the bridge authority, will be available for use for time to come.

Perhaps even more important than maintenance to drivers using the bridges is the low tolls afforded by the New York State Bridge Authority. In a time when a cup of fancy coffee costs a few dollars, the bridge toll is less than a cup of coffee and provides peace of mind that the bridge you are crossing is safe. The bridge authority has always strived to provide bridge crossings with the lowest possible tolls. When it was formed in 1932 and took possession of the Mid-Hudson Bridge in 1933, the toll for an automobile and just the driver one way was 80¢! Each passenger was an additional 10¢. Animals ranged from 10 to 30¢, and pedestrians paid 10¢ to walk across the bridge one way. The bridge authority, which consisted of just three members at that time, had no employees. They had to meet in the office of Col. Frederick Stuart Greene, the superintendent of the New York State Department of Public Works. Through dogged determination, that first bridge authority board fought the Department of Public Works, taking that fight all the way to the US Congress, to get the tolls lowered to a reasonable amount so the average person could actually use the bridges. Finally in 1937, the toll was reduced to 50¢ each way, with no additional charge for passengers in the automobile. Pedestrians were still charged to walk across the bridge for another few decades, but the toll became much more reasonable for motorists. When the bridge authority acquired the Bear Mountain Bridge in 1940, the first thing it did was reduce the tolls on the Bear Mountain to the same as the other bridge tolls. Tolls were reduced again during World War II to as low as 25¢. The bridge authority is required by law to charge the same toll on all its bridges, thus preventing the need for driving out of your way to cross a bridge with a less expensive toll. By 1970, to reduce traffic at the toll booths, tolls were collected in the eastbound direction only.

In many ways, these bridges were feats of engineering—the firsts of their time. World-renowned engineers, like Ralph Modjeski and D. B. Steinman, designed and oversaw the building of these bridges. Most motorists drive across these structures with no idea of the history behind them. The hope is that this book will help provide some of the amazing history of these bridges we cross every day, providing an opportunity to appreciate not just the beautiful views but also the engineering and the caring maintenance of these iconic structures by the New York State Bridge Authority.

One

MID-HUDSON BRIDGE

Across the river from the bustle of Poughkeepsie in the early 1920s was Highland Landing, a quiet river community with a local train stop. Beyond Highland Landing to points west were orchards and farms with produce enough for the whole country. Trucks, the new transport vehicle for produce, had to drive to Albany to reach other markets. The approach to a vehicular bridge would replace the landing, creating an opportunity for growth on both sides of the Hudson.

On June 2, 1923, an act of the New York State Legislature provided that a vehicular bridge crossing the Hudson River shall be located within the city limits of Poughkeepsie. Modjeski and Moran were appointed engineers for the project 17 days later. Preliminary riverbed borings and studies of the shore access points determined the best location for a suspension bridge. Much work was done to clear and grade city streets and provide access to the approaches.

With the plans approved, the funds procured from the state, and preliminary work begun on approaches to the bridge, the cornerstone setting in Poughkeepsie was celebrated by thousands and witnessed by then Gov. Al Smith on October 14, 1925. It would be almost five years before the bridge was completed and opened to the public. Before work could begin in the river, the approaches and anchorages were begun.

The New York Central Railroad runs the length of the Hudson River from New York City north to Albany along the east shore. The east approach of the Mid-Hudson Bridge will span the tracks with an arched truss bridge. Care is taken to not interrupt the train traffic, as will be the case building all Hudson River bridges to come. The east abutment for the bridge crossing the tracks is seen nestled up against buildings that will soon be removed. Seen in the photograph above, the two smaller white block sections in front of the abutment are skewbacks that will support the end structure of the bridge arch. In the image below, the end of the arch sits on the skewbacks as the truss sections are completed.

An area of Kaal Rock, a natural rock outcropping, was chosen for the east anchorage of the Mid-Hudson Bridge. Seen here in December 1925, the anchorage is being dug to be ready once the approaches reach the river and caissons are sunk, while pier and towers stand waiting for the main cable to be spun.

As work progresses on the roadbed of the approach bridge crossing the New York Central Railroad tracks, caissons are being fabricated in the yards of the Staten Island Shipping Company and towed to form the base of the piers. Three bridges carried the approach, completed by Scott Brothers Construction Company in the summer of 1926, over city streets and the railroad tracks.

Prior to the use of concrete mixer trucks, portable concrete batching plants made it possible to create roadways. Portland cement, named by its English inventor for its resemblance to high-quality building stones found in Portland, England, was likely used in the batching plant seen here. Additives such as gypsum, possibly in the bags stacked to the right, were used to control setting properties.

The west anchor pits were dug into the rock bluff on the west shore, providing a natural anchorage for the cables and saving the expense of a massive and costly masonry anchorage. The rock bluff would also support the abutment and the perpendicular approach road to the bridge. Initial plans called for a south and north T approach, but only a northerly turn off the bridge was constructed.

The east pier caisson reached Poughkeepsie on March 26, 1927; the west pier caisson arrived on April 23, 1927. Open caissons were used to construct the Mid-Hudson Bridge piers. As the caisson was sunk, aided by a concrete false bottom, walls were built up to extend to the height above the water. Once on the soft river bottom, the cutting edge was used to break through to firmer footing. Dredging removed earth from the bottom of the caisson.

While the west caisson proceeded without incident, the east caisson provided what would become an epic feat of engineering. At 4:30 a.m. on July 27, the caisson listed to an angle of 42.5 degrees from the vertical in less than a minute. The east edge dropped 29 feet; the west edge rose 11 feet. Steps were taken to keep the caisson from tipping further, and the righting process began. Pictured here in October 1927, very little progress was seen until the spring.

Although it was widely understood that if the east caisson could not be righted, the bridge could not be built, work continued on the shore. The east abutment has been built with the saddle blocks that will hold the cables on the sides of the top. The cables will cross the saddle blocks and bend towards the eyebars, where they will be fastened in the anchorage.

Bridge towers and trusses, fabricated in Ambridge, Pennsylvania, for the American Bridge Company, were painted and shipped in the summer of 1927. They are seen here resting on the Poughkeepsie waterfront, waiting to be installed. The towers were prebuilt in three sections with field-riveted vertical joints to facilitate quicker construction. The tipped caisson has extended their wait.

By May 1928, a slight righting motion was noted. Attempts to right the caisson included a system of pulleys, counterweights, and river-bottom excavation, all of which required divers to provide assistance. A "carfloat," as the device on the right is called, enabled the lines to remain constantly taut, even with the changing tide. A self-propelled jet, developed by the Navy, proved most helpful with excavation.

The west pier, in the foreground, was completed on June 12, 1928. Work continued to right the east caisson, which, once begun, proceeded at a rate of 30 minutes of arc, or one half of a degree, daily. At that rate, the 42.5-degree list of the caisson would take about three months to become vertical.

By early August, the caisson had very nearly been brought to the vertical. The righting appliances were removed. Attempts were made to ensure the caisson landed as close to the originally intended position as possible. Although Blakeslee Rollins Corporation, the contractor for the piers, had very nearly escaped disaster, they were far from done.

On November 20, 1928, the final sinking of the caisson was started with the cutting edge at a depth of 112 feet below water. It took 18 days to sink to the final depth of 134.4 feet, resting on a stratum of coarse gravel. By December 26, 1928, about 13,000 cubic yards of concrete had been deposited in the upstream half of the caisson. Operations were closed for the winter.

Work resumed on March 19, 1929. A cofferdam was erected to provide a dry area enabling the installation of the granite block outer section of the pier. Granite would ensure a firm surface, almost invulnerable in the swirling force of the waters of the Hudson River as it flows south with the movement of the tides flowing north.

Completed on May 16, 1929, the last stone was ceremonially placed on May 22, 1929, at the end of a herculean effort. Blakeslee Rollins Corporation, the substructure contractor, sued to receive compensation for the additional work and time spent to right the caisson. In the end, it was determined by a court that it was not a design issue but the fault of the contractor that had caused the caisson to tip.

Once the piers were complete, the towers could be assembled. The west tower, seen here in June 1929, was completed and waiting for the east tower to catch up. The towers were erected by a crew of 35 men, including those at the base where the steel was delivered on barges. The derrick boat *Colossus* used its 125-foot boom to lift the lower sections in place. A crawler traveler positioned on the tower erected the rest.

Less than one year from the bridge opening, the towers are complete. The west abutment is in place, seen here in the bottom right. Once the rocker saddles are in place, footwalks and then the main cables can be strung. Each bridge tower stands 315 feet above the water's surface and is 4,000 tons of steel, with 52,000 rivets used per tower.

Barges are ready to begin laying the cable for the footwalk. The two footwalks will provide a platform for the workers stringing the two main cables. Once the cables are in place, the floorboards will be installed. The barge, pushed by tugs, contains four spools of wire. It will cross the river laying the four cables for each footwalk at the same time. Those cables will then be hoisted into place.

Nearly complete footwalks provide a platform for workers. Follow the footwalks into Poughkeepsie, past the abutment, to the approach road. At the end of the approach is Union Square. A caution sign, located on the riverbank to the right below, marks where a submarine cable was laid from the east shore to the east pier supplying 220-volt AC power to construction equipment; a similar cable was laid on the west side.

Seen clearly against the light-colored block of the exterior of the west abutment, a worker stands on the top of the wall looking down towards the cable spinning equipment ready to begin spinning the main cables. Footwalks, each a simple suspension footbridge with half-inch wire rope handrails and a wooden platform nine feet wide, are supported on two pairs of one-and-seven-eighths-inch steel suspension cables, with four cables for each footwalk. The footwalk cables are each 3,285 feet long and weigh in total about 10 tons. Delivered by boat and hauled up from the river by a hoisting engine, the ends of the footwalk cables will be temporarily attached to eyebars in the anchorage. Each main cable will not rest on the footwalk but will hang one foot, ten inches above it as individual cable wires are spun from anchorage to anchorage until the cable reaches its completed size.

Spun from anchorage to anchorage over the top of each tower, the main cables are placed in their permanent position in the saddle atop each tower. Each strand, or cluster of wires, seen here contains 320 parts of a single continuous wire with a loop at each end that is attached with a special connection to an anchorage eyebar.

Spinning wheels about five feet in diameter, held here by the worker in the southeast anchorage, were attached to hauling ropes that pulled the cable wires from anchorage to tower tops, where they were disconnected from the wire, moved to the other side of the tower, reconnected, pulled across to the other tower, and ended at the opposite anchorage. Each wheel would make that round trip in about 18 minutes.

Each cluster of 320 wires was compacted by workers with special hand tongs to a diameter of three and seven-eighths inches, and then the compacted strands were seized or encircled with tin bands. Observers directed any necessary adjustments to wires from their positions at the center of the main and side bridge spans.

After all strands were completed, each cable was compacted to a diameter of 16¾ inches. Four compacting machines operated by 30 men, using three hydraulic jacks, forced all wire strands into complete and continuous contact. Spaces between strands were eliminated, resulting in the permanent parallel position of all wires. Additional bands, seen on the footwalk, were wrapped at three-foot intervals to hold the compacted cable prior to final wrapping.

The completed main cable enters the east anchorage, where the clusters of wires are attached to the eyebars that are attached to more eyebars that are embedded into the rock below. A group of workers can be seen banding the cable just above where the clusters split off to the eyebars. Once completed, the base level of the 500 tons of eyebars in all four anchorages were cemented permanently in the anchorages.

The 34,000 feet of suspender ropes, or cables, were fabricated at the Worcester Plant of the American Steel and Wire Company. After suspender ropes were cut to length, both ends were unwrapped for about seven and a half inches to enable them to be placed in a special clamping machine, seen here in the shed, which connected them to the socket.

Truss and floor erection began at the towers, as seen here. Materials were hoisted into place from barges in the river. Suspender cables, seen hanging from the main cable, are attached to cable bands and will hold up the deck once attached to the side stiffening trusses. Traveler cranes attached to the floor moved along as the next section was erected from the towers out to the abutments.

Truss erection can be seen here as the floor system approaches the west abutment. The stiffening truss verticals, often referred to as "pork chops" for their similar shape, are attached to much shorter suspender ropes as the main cable dips down to the anchorage. The traveler crane is the triangular erection device attached to the temporary supports above the floor system.

The truss erection is nearly complete, with less than three months to go until opening day. In this image from just below the east tower, the length of the bridge can be seen crossing the river to the west shore. Suspender ropes wrap the main cable at each cable band, then reach down through a space in the footwalk to the stiffening truss holding up the floor system. With the assembly of the truss diagonals, chords, and sidewalk steel, a maximum force of 160 men finished the erection of all three spans in about 40 days. In total, there were about 200,000 steel rivets driven by 12 four-man gangs in about 42 days. During the erection of the steelwork, rigid inspection was carefully maintained to ensure accuracy and high quality of workmanship, and ample precautions were taken for the safety of all workers. At all hours when work was not in active progress, a regular inspection patrol was maintained over the entire structure.

Less than a month from opening day, the steel structure of the bridge, seen here from the east span through the main span to the west tower, is 3,000 feet long between anchorages. It has a main span of 1,495 feet, with 750 feet of the west span and 755 feet of the east span, due to repositioning of the east tower after the righting of the tipped caisson. The roadway width is 30 feet, and the sidewalks are located outside the stiffening trusses. The structure contains 9,450 tons of structural steel, 2,200 tons of cable wire and suspender ropes, and 750 tons of castings and railings. There are 282 suspender cables hung from 16¾-inch-diameter main cables. The Mid-Hudson Bridge would be one of the longest suspension bridges in the country when it opened on August 25, 1930.

This cable wrapping machine was power-driven, wrapping 80 feet of cable per day. Parts of the cable were hand wrapped, progressing much slower at about 6 feet per day. No. 9 galvanized steel wire was used to wrap the cable. All cables were painted with one coat of red lead, then two coats of lead and oil paint. The lead paint was abated in 2000, when the bridge was repainted.

Scott Brothers Construction Company of Rome, New York, poured the roadway and sidewalks. The bridge would have two traffic lanes, one in each direction. That was sufficient in 1930 but would require refiguring years later as bridge traffic increased. The remedy was to create a third, narrower center lane on the bridge during high traffic times to allow for easterly traffic flow in the mornings and westerly flow in the evenings.

About 15,000 gathered in Poughkeepsie for the opening of the Mid-Hudson Bridge on Monday, August 25, 1930. A parade in Poughkeepsie began festivities, followed by speeches in Union Square before the car carrying Governor Roosevelt would cross the newly opened bridge. Gathered in the crowd were 1,000 children, assembled as a "court of honor" for Governor Roosevelt and former governor Smith. Baskets of flowers from the city's greenhouses were presented to Eleanor Roosevelt and Catherine Smith. (Modjeski & Masters Company Records, Archives Center, NMAH, Smithsonian.)

Catherine Smith, in the center of the back seat, cut the ribbon on the east side of the bridge. The official car then proceeded to cross the bridge with Gov. Franklin Roosevelt in the back with Mrs. Smith and Eleanor Roosevelt. Col. Frederick Stuart Greene stands at the front of the far side of the car. Gov. Al Smith is seated in the front. Governor Roosevelt remarked that his forebears had come to this area 200 years ago, and it was about high time he visited his cousins in Ulster County. (FDR Presidential Library & Museum.)

Once across the bridge, Eleanor Roosevelt cut the ribbon. On the left is Colonel Greene, who had directed the construction, and to the right are Ralph Modjeski and Daniel E. Moran (far right), engineers who designed the bridge. The governors made brief statements, calling the new bridge a great connector not just for New York but for the entire Union. Then the car proceeded into Highland before turning around and heading back to Poughkeepsie.

Once the official cars had exited the bridge, pedestrians were allowed access. At the east approach, 10,000 people jammed in the crowd waiting to cross. First across were a hundred small boys who had elbowed their way to the front of the crowd. They ran up the approach yelling and cheering. Crowds came from Highland, all enjoying the beautiful structure and views. (Modjeski & Masters Records, Archives Center, NMAH, Smithsonian.)

Celebrating the opening of the Mid-Hudson Bridge, Highland would not be outdone, having a great community celebration lasting from early afternoon until late at night. Festivities began with a grand parade complete with many bands, multiple fire companies, Scouts and area clubs, and community officials. Suppers were served free of charge by area churches to over 500 residents. To end the day, a dance was held in Smith Hall. (Town of Lloyd Historian.)

The east entrance, four days after the opening, was a much calmer place. The toll collector's quarters, the stone structure at right, was built to provide collectors a place to stay, but little record remains of its use by collectors. On opening day, Governor Roosevelt proclaimed the bridge be named the Mid-Hudson Bridge. In 1994, by an act of the legislature, the bridge was renamed the Franklin D. Roosevelt Mid-Hudson Bridge.

Many improvements would follow in the ensuing years. New toll booths on the east side are seen here in 1958. Tolls when the bridge opened in 1930 comprised a complicated structure. Passenger cars were charged 80¢, and additional passengers were 10¢ each. Pedestrians and bicycles were 10¢. Animals of various sizes ranged from 10 to 30¢. All tolls were charged in both directions.

The original approach on the west side of the bridge required a 90-degree right turn immediately upon exiting the bridge. Straight ahead of the exit was a rock cliff. A new approach, opened in 1967, required blasting through rock but provided a better turn up the hill to a new row of toll booths. By 1970, tolls on all the bridge authority bridges were collected only when traveling east.

In an effort to share this beautiful structure by providing better public access to the bridge sidewalk, a new pedestrian walkway was opened in 1999. In the summer of 2001, necklace lighting was installed, adding to the iconic nature of the Mid-Hudson Bridge. Greg Herd, manager of information technologies, designed a system allowing light shows to be scheduled to run at specific times for celebrating events and marking special occasions.

Composer Joseph Bertolozzi draws sound from the cross-frame below the deck of the Mid-Hudson Bridge for Bridge Music, a unique sound-art installation featuring the bridge itself as the instrument. Bertolozzi drummed and scraped girders, cables, and more with various mallets, then composed music using only those sounds. It can be heard at listening stations on the bridge's towers and 95.3 FM in the parks flanking the bridge. (Photograph by Blue Wings Press.)

The bridge authority has been considered a family by employees over the years. There is no better evidence of that than the two people here, John Brooks (left) and his dad, Cecil. Cecil began at the bridge authority in 1957, retiring in 1977. John began in 1966, retiring from full time in 1997 but continuing part-time well into his 80s. A Brooks has worked at the authority for over 63 years, making it the "New York State Brooks Authority."

Executive director Harry Stanton, approachable and well-liked by all staff and interested in the work and well-being of each employee, often visited each bridge site. He appeared at construction worksites, grabbed a flag from an employee to direct traffic, or stepped in a tollbooth to collect tolls. He truly led by example, garnering respect and admiration from the entire authority "family." Stanton is the only executive director to sit on the bridge board.

Two

Rip Van Winkle Bridge

Created on March 31, 1932, with the signing of a bill by Governor Roosevelt, the New York State Bridge Authority would finance the building of the Rip Van Winkle Bridge and, on March 17, 1933, assume the ownership and management of the Mid-Hudson Bridge. Designed by Glenn B. Woodruff, principal engineer of the Mid-Hudson Bridge, work began on the Rip Van Winkle Bridge on April 1, 1933. (Rip Van Winkle Bridge Photographs, Archives Center, NMAH, Smithsonian.)

Frederick Snare Corporation was awarded the substructure contract for the Rip Van Winkle Bridge when it outbid 11 other companies, one from as far away as Iowa. Work was scarce during the Depression. The bridge authority would require contractors constructing the bridge to hire a large percentage of local labor. In addition, materials available locally in Greene and Columbia Counties, like cement, must be used to further assist local businesses. Snare would construct the bridge's concrete foundations and piers, then the masonry tower pedestals. Four piers are located in the main channel of the Hudson River, with 10 crossing Roger's Island to the east abutment. Work on the first water pier on the west is seen in the photograph above. Below is a batching plant on a barge. It combines the necessary ingredients, enabling the concrete to be poured in any location, including into water piers. (Both, Rip Van Winkle Bridge Photographs, Archives Center, NMAH, Smithsonian.)

Frederick Snare Corporation utilized a temporary narrow-gauge railroad to move materials while constructing the piers on Rogers Island. The piers seen here, in April 1934, are supported on pipe piles driven to rock. The pipes used were 14-inch diameter, with the water removed and the pipes filled with cement before the pier was built on top. (Rip Van Winkle Bridge Photographs, Archives Center, NMAH, Smithsonian.)

Harris Structural Steel Company of New Jersey supplied and erected the steel superstructure of the Rip Van Winkle Bridge. The steel structure was erected by using a traveling crane that hoisted steel members from the ground or from barges on the water. Erection would begin with the bridge towers, and then using falsework, or temporary piers, the cantilevered truss sections were supported until they were connected to the next section. (Rip Van Winkle Bridge Photographs, Archives Center, NMAH, Smithsonian.)

Completed cantilever truss sections of the bridge are seen above on August 7, 1934. The traveling crane can be seen moving along as it erects more steel. The bridge construction has reached the main channel, where the remaining piers are in the river. With a total length of 5,041 feet, the main channel, an 800-foot-long cantilever, suspended truss section, is yet to be spanned. (Rip Van Winkle Bridge Photographs, Archives Center, NMAH, Smithsonian.)

Looking southeast towards the bridge structure, Olana, home to Frederic Church, can be seen on top of the hill above the east end of the bridge. Church was a student of Thomas Cole, a Hudson River painter, whose home was on the west side of the Hudson River. With pier 2 in the foreground, erection has begun on the through cantilever span over the river channel. (Rip Van Winkle Bridge Photographs, Archives Center, NMAH, Smithsonian.)

The west approach to the bridge was originally planned to cross the Thomas Cole property. Descendants of Thomas Cole, founder of the Hudson River painters, thought his property should be worth much more than the bridge authority was prepared to pay, so the approach was moved a bit north and, today, passes right by the Thomas Cole National Historic Site. (Rip Van Winkle Bridge Photographs, Archives Center, NMAH, Smithsonian.)

With foundation work completed, Frederick Snare Corporation turned its focus to completion of the Rip Van Winkle Bridge administration building. Built in Dutch Colonial style, the building would become permanent offices for the New York State Bridge Authority, previously located in the old Worth Hotel in Hudson, New York, across the river. Harris would go on to erect the steel structure of the Kingston-Rhinecliff Bridge and, later, the Castleton-on-the-Hudson Bridge. (Rip Van Winkle Bridge Photographs, Archives Center, NMAH, Smithsonian.)

Atop the almost completed administration building sits a weather vane hosting old Rip Van Winkle himself. The weather vane is still in place today. Another reminder of Rip Van Winkle is a very large boulder located in the park outside the building, excavated at the site; legend has it that Rip slept there. (Rip Van Winkle Bridge Photographs, Archives Center, NMAH, Smithsonian.)

When the steel arrived at the bridge site, it was painted a protective red. Workers painted it a silver color before it could be lifted into position and hand riveted in place. Employing nearly 750 men at steel mills and the bridge construction site for two years when unemployment was at record highs, the New York State Bridge Authority was helping ensure locals had work. (Rip Van Winkle Bridge Photographs, Archives Center, NMAH, Smithsonian.)

Looking northwest from a boat in the Hudson just southeast of pier 3, on December 4, 1934, four falseworks support the anchor through-truss section that will be the force to support the suspended through-truss cantilevered center section of the bridge. Ice forming on the river would make it necessary to erect the remaining span over the main river channel without falsework, working from the top of the structure. (Rip Van Winkle Bridge Photographs, Archives Center, NMAH, Smithsonian.)

Working both ends towards the center of the suspended span, hydraulic jacks force the last sections into place in January 1935. Driving in the final rivets, the main shipping channel of the Hudson River had been spanned. The navigable part of the channel has a width of 765 feet, the clearance to the bridge from the water is 142 feet. The bridge would open less than six months later. (Rip Van Winkle Bridge Photographs, Archives Center, NMAH, Smithsonian.)

The Rip Van Winkle is unique for a Hudson River bridge in that while it bridges the Hudson, it crosses almost as much land as water. Looking northwest from just southeast of the east abutment, the full length of the viaduct can be seen. The underdeck truss of most of the bridge length will allow for unobstructed views of the "Birthplace of American Art." (Rip Van Winkle Bridge Photographs, Archives Center, NMAH, Smithsonian.)

Workers are setting up to lay in the concrete on May 16, 1935. Corbetta Construction Company was contracted to pave the bridge deck and the approaches. One of the nation's most successful construction firms, Corbetta—or "Mr. Concrete," as the company owner was nicknamed—pioneered techniques like using plywood for concrete forms, concrete pumping, and site pre-casting. (Rip Van Winkle Bridge Photographs, Archives Center, NMAH, Smithsonian.)

Deck concreting was accomplished by a combination of equipment seen here. A standard paver discharged the concrete onto an inclined belt conveyor, which dumped the concrete into twin hoppers of a duplex concrete pump. The three pieces of equipment were coupled together and moved forward as a unit. Paving began at the east abutment and proceeded westward at a rate of 250 feet of roadway deck per day. (Rip Van Winkle Bridge Photographs, Archives Center, NMAH, Smithsonian.)

Workers seen here smoothing out the concrete earned $1 per hour for a maximum of 30 hours per week. A brief strike during construction allowed for workers to work additional hours. The finished roadway between curbs measured 30 feet across. A sidewalk on the south side was a bit over four feet wide. Metal railings ran the full length of the bridge. (Rip Van Winkle Bridge Photographs, Archives Center, NMAH, Smithsonian.)

The opening and dedication of the Rip Van Winkle Bridge on July 2, 1935, coincided with the weeklong festivities in Hudson, New York, to mark the 150th anniversary of that city. Gov. Herbert H. Lehman and Col. Frederick Stuart Greene, superintendent of public works as well as supervisor of the bridge construction, joined other dignitaries on the Catskill side of the bridge.

Following a parade through Catskill and speeches by dignitaries, the crowd awaits the cutting of the ribbon on the west side of the bridge in front of the bunting-covered administration building. Grace Greene, wife of Col. F.S. Greene, will cut the west side ribbon, with Laura Miller, daughter of bridge authority board president Clifford Miller, cutting the ribbon on the east side of the bridge. (Rip Van Winkle Bridge Photographs, Archives Center, NMAH, Smithsonian.)

Following the opening motorcade across the bridge, vehicles and pedestrians were permitted to cross the bridge free of toll until 6:00 p.m. that evening, when the first toll was collected. More than 3,000 vehicles took advantage of the free trip. The initial toll structure was the same as the tolls on the five-year-old Mid-Hudson Bridge, but the bridge authority would reduce the toll on both bridges to 50¢ in 1937.

The Rip Van Winkle Bridge, seen here in later years, provides a beautiful view from the new sidewalk, built in 1996. Cantilevered off the south side of the bridge, it provides for a safer walk and a wider roadway. Many locals and visitors to the area enjoy the park site, beautifully maintained by the bridge authority, and the nearly mile-long walk across the bridge.

Officially opened on June 1, 2019, the Hudson River Skywalk was a community effort to bridge between Thomas Cole National Historic Site and Olana State Historic Site by way of the Rip Van Winkle Bridge and a path connecting the two sites. Seen here in front of the podium at the opening ceremony is Tara Sullivan, the driving force behind the project, and behind the podium are Sean Sawyer, from Olana, and Betsy Jacks, from Thomas Cole.

A large contingent from Greene County, carrying green flags, met a large contingent from Columbia County, carrying blue flags, at the center bump-out in the bridge sidewalk. Built by the bridge authority for the purpose of enjoying the spectacular views that inspired the Hudson River painters, there are three viewing decks along the walk, providing space to stop and enjoy the view or set up an easel and paint.

The incredible park on the west side of the Rip Van Winkle Bridge is being used here for a STEAM workshop for New York State Master Teachers. In the foreground, educators are learning how to represent the design of the bridge artistically. Under the two canopies in the background, educators from the Olana and Thomas Cole sites are working with Master Teachers to explain the important connection between art and science.

The Skywalk Arts Festival, an annual event begun in September 2017, has become very popular with local artists and art lovers. An opportunity to showcase local art, the focus is landscape painting and photography of the Hudson River, the Hudson River valley, and of course, the Hudson River bridges. Artists set up in the Rip Van Winkle Bridge park area under canopies overlooking the Hudson River.

A favorite activity among visitors to the Skywalk Arts Festival is trying their hand at creating landscape art. Some budding young artists are learning from art educator Stasia Fernandez, top left, how to use watercolor pastels to create a beautiful painting of the Rip Van Winkle Bridge. As they paint, they are learning to appreciate the beauty around them and their ability to represent it artistically.

The original fog bell, once mounted under the Rip Van Winkle Bridge, has been refurbished. Marc Griffin (left) proposed the bell be rehung and made functional again. With funding from Historic Bridges of the Hudson Valley, Verdin Company of Cincinnati refurbished the bell, while Julie Truncali, an engineering intern, designed its new home. The structure was fabricated in house and installed by Jack Kelly (right) and his maintenance crew.

Three

Bear Mountain Bridge

The Bear Mountain Bridge, built by the Bear Mountain Hudson River Bridge Company, was the first vehicular crossing of the Hudson River south of Albany. When it opened on November 27, 1924, it was the longest suspension bridge ever built. The bridge company, a private entity funded by stockholders and organized by its president, E. Roland Harriman, was able to build the bridge by a special act of the state legislature in February 1922.

The two bridge towers, unlike other Hudson River bridges to come, were located on the shore. The creeper derrick, on top of the structure, moved up the tower as it hoisted sections in place. Since the mid-1800s, the location was the choice spot to build a crossing, as it is the narrowest part of the river. However, no bridge was built until the Bear Mountain Bridge, since Anthony's Nose, on the east, needed to be tunneled as the mountain had no level access for trains. Access for vehicles would require blasting out a road from the side of the mountain. Construction of the "goat trail" would cost almost as much as the bridge. Jackhammers for breaking up anchorage rock worked by using compressors, seen below, for power. Once broken up, workers dug out and hauled away rock debris.

Terry and Tench, Inc., of New York City was contracted to construct the bridge on March 24, 1923. Construction began in May 1923. The anchorage pits were hand-dug, with the help of jackhammers to break the rock, to depths of 90 feet for the east pits and 110 feet for the west pits. The towers, begun in October, rested on steel bases 10 feet by 35 feet by 5 feet high, which rested on concrete piers 20 by 20 feet. Each tower was 351 feet high, with the roadway running through a semi-circular arch in the tower about halfway up from the river. Underdeck truss members, seen here, are being erected on the approach spans of the bridge. A traveler, or moving crane, was used to hoist the steel from barges on the river into place on the structure. Riveters, working in pairs, would follow, permanently connecting the steel. The white piers, bottom center and left, will support the side spans of the bridge. The Bear Mountain Bridge was designed to have only the center span suspended.

To enable the spinning of the parallel wires of the main cables for the bridge, a temporary work platform, or footbridge, must be erected directly below and about three feet from the main cable. The barge (bottom right) carries four spools that will lay four cables for each footbridge to be hoisted up between the towers. Cross bridges, connecting the two long footbridges, will allow workers to go between the main cables and will also serve to stiffen the footbridges laterally to protect from strong winds. Once the cables are in place, the floor system for each footbridge, assembled in sections on the ground, will be hoisted in place. It was imperative to hoist sections on each side of the bridge at the same time to ensure balance of the footbridge. Both footbridges were in place by early June.

Roebling cable spools, bottom right, will be placed in the spinning shed at the end of the footbridge, as cable wire is needed to spin the main cable. The footbridges, seen here on May 23, 1924, were complete, and main cable spinning began. Extensive in-factory testing was completed by Roebling to ensure the success of the new parallel wire cable construction to be used on the Bear Mountain Bridge.

Six workers standing together on the footbridge give good evidence of its width. Rope handrails enhance the safety of the platform. The first four cables, spun together from anchorage to anchorage, can be seen here. Workers are adjusting the guide ropes that will help ensure uniform, accurate length of individual wires as they are spun to make each main cable.

The west anchorage, seen here, consists of the anchorage pit (bottom left) and the spinning shed (on the platform at the top). More Roebling cable spools are on hand, each containing approximately 85,000 feet of continuous-length wire. Ends of the coils are split to enable them to be spliced to the next coil to ensure continuous cable from anchorage to anchorage. That anchorage is seen here upon completion of the south cable.

The cable has been compacted, held together by metal bands, and then split into strands at the metal casting that holds the cable wire in place. Each cable has 37 strands, or clustered cable, made up of 196 individual wires. The anchorage is filled with cement, poured from above the rock opening, permanently holding it in the ground.

By September 1924, the main cables were almost complete. Here on the west side of the bridge, the south cable, right, is complete, while work continues on the north cable. Apparatus for spinning the wires is clearly shown above the north, or left, footbridge, where what looks like three doorframes stand above the footbridge. Cable wire is pulled through those frameworks, using a tramway and a spinning wheel, four wires at a time, up and over each tower and down to the opposite anchorage, where the wires are attached to the casting in the anchorage. Then four new wires are sent back to the other anchorage using the same process. All cable wire used in fabricating the two main cables, and the steel wire rope used for footbridges and suspender ropes, were furnished by John A. Roebling's Sons Company, of Trenton, New Jersey.

In the photograph above, the main cable (bottom left) can be seen running below the construction of the administration building on the west side of the bridge. The anchorage is visible from the basement of the building, where the cable reaches the cemented pit well below the building. A cement mixer is providing material for the building's foundation. The full length of the main span can be seen in the image to the left. The backstays, the main cable from tower to anchorage, are unloaded. The backstays have no suspender ropes. Only the main span between the towers is suspended. The side spans are supported by the underdeck trusses. This is evident in the curvature of the main span cables between the towers, while the backstays are straight.

Suspender ropes, or cables, are hung over the main cable with the two ends reaching down to hold up the deck. At the main cable, the suspender ropes are held in the cable band. The ends of the suspender are sealed into a connector that attaches to the truss struts lifted from barges to form the bridge decking.

Concreting of the road surface began on the eastern approach while work continued to complete the decking of the main span. Stone walls framing the bridge entrance from the north and the south will give a finished look to the roadway. In the distance, the Popolopen Creek flows into the Hudson. The arch bridge carrying Route 9W from Highland Falls to Bear Mountain stands high above the creek.

With the barge in the river supplying the remaining steel trusses, work continues to connect the bottom chord at the center of the main span. Section after section has been placed until the superstructure is almost complete. The travelers move closer together until the two lower chords create a solid connection from shore to shore.

The roadway of the main span is seen here from the west tower across to Anthony's Nose. Concrete forms are placed on the deck, ready to be filled. Work continues on the highway at the southeast entrance to the bridge. A toll booth was constructed at the beginning of that roadway on the other side of Anthony's Nose. Tolls collected will pay for insurance for accidents on that precarious road.

Much activity can be seen by mid-October on the west span of the bridge. Railings have been installed, and sidewalks on either side have forms for concrete. A large group of men and women are gathered in the lower left of the photograph. The bridge will open in a little more than a month. Main cables are compacted and wrapped from anchorage to anchorage.

On the east side of the bridge, work is nearing completion. The "goat trail" cuts through the side of Anthony's Nose and provides travelers with incredible views of the Hudson River and the Highlands. A few overlooks were constructed to safely park cars and enjoy the view. The steam of an engine can be seen entering the tunnel that runs under the east end of the bridge.

Looking west from the south approach road, the main cable dips below the roadway at the center of the span. The design allows an unobstructed view of the river. The curved arch in the tower above the roadway mimics the entrance to a tunnel. Concerns that a vehicular bridge would mar the scenery could be put to rest. The Bear Mountain Bridge stands proudly as a beautiful example of engineering.

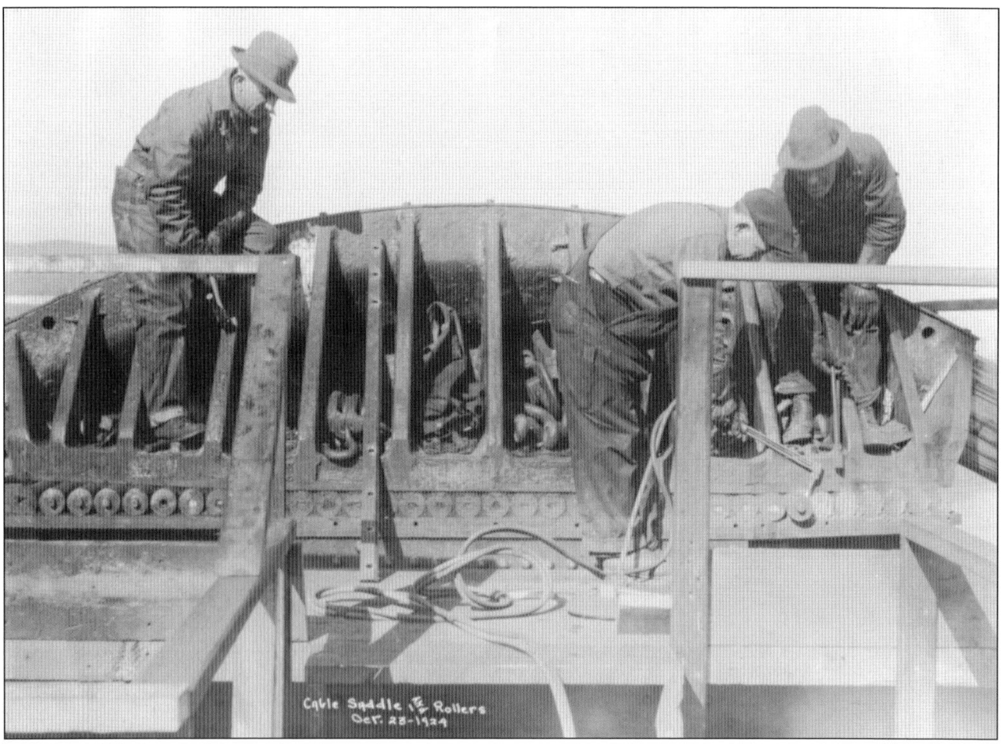

Tower saddles were bolted firmly to the tower top once the superstructure was completed and the total weight of the finished roadway was suspended from the cables. To prevent possible rotation of the cables, metal cement was applied to keep them solidly in place. This is Roebling's first use of parallel wire cables, making the Bear Mountain Bridge an engineering first.

With the majority of the bridge complete, the footbridges will be dismantled and removed. The administration building is still under construction and would not be completed in time for the bridge opening. The small structure to the left serves as a toll booth. The bridge has been constructed in record time, although work on the "goat trail" has cost the contractor much more than anticipated.

On November 26, 1924, one day before it opened to traffic on Thanksgiving, a motorcade of cars left the Peekskill station. Traveling the east approach road and stopping at the end of the bridge, they witnessed the ribbon cutting. Driving four cars abreast, they were the first automobiles to cross the bridge. Mary Harriman, widow of E.H. Harriman, pulled the cord to unveil the bronze tablet dedicated to all who built the bridge.

Tolls were expensive on the new bridge. At a time when a new Ford cost $528, the average family of four would pay $1.15 in tolls, each way, to visit the Bear Mountain Park. It cost 20¢ just to walk across the bridge. Many automobiles parked at the side of the bridge to enjoy the view. With tolls that expensive, the bridge itself was a destination.

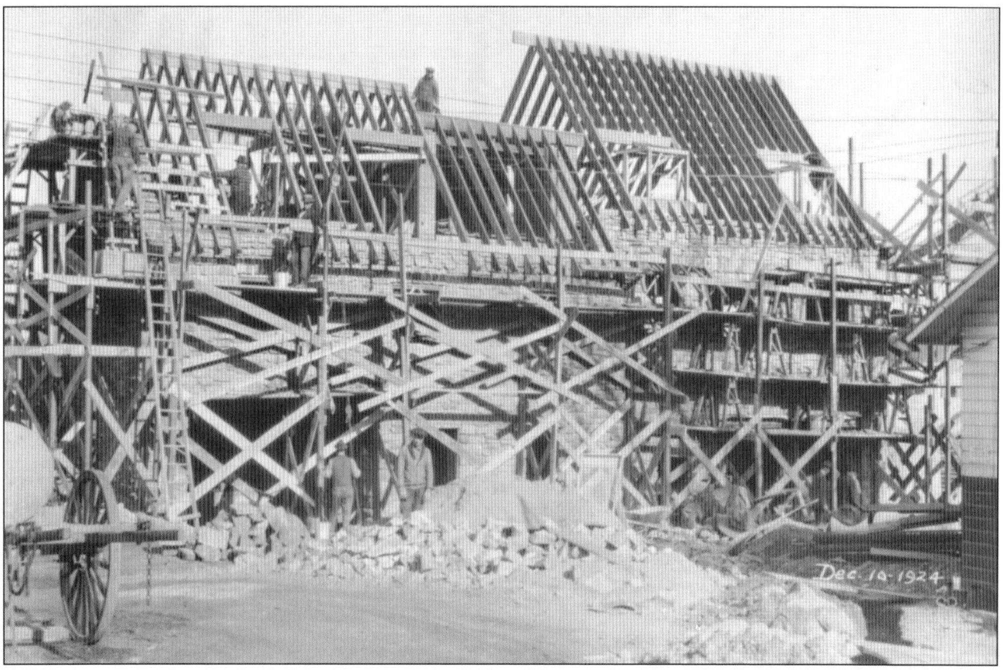

Work continued on the bridge administration building, seen here on December 10, 1924. Once completed, the bridge manager would take up residence. It is said that the first bridge manager stayed at the bridge every day, all day, sleeping there at night. He is said to have left once during his tenure to give his daughter in marriage. Once done, he returned to the bridge, and the wedding celebration continued without him.

As the automobile approaches the Bear Mountain Bridge, it passes the sign on the right that reads, "Caution: Toll Bridge and Toll Road 500 feet ahead." The toll was collected on the west side of the bridge. Should the driver decide to go south, a toll was collected on the bridge highway on the way to Peekskill. The Bear Mountain Bridge, although beautiful, was a financial disaster.

As part of the original agreement between New York State and the Bear Mountain Hudson River Bridge Company, after a certain period of time, the bridge could be turned over to the state. Paying the purchase price of $2,275,000 to pay off the bonds and stock, the New York State Bridge Authority took possession of the bridge, immediately reducing the toll to 50¢.

In 2024, the Bear Mountain Bridge turns 100 years old. The bridge authority carefully maintains this iconic structure. Suspension bridges require special care. The Bear Mountain and the Mid-Hudson Bridges have their main cables inspected at regular intervals. Inspectors have unwrapped the cable and are looking for wear. Jeff Wright, chief engineer at the bridge authority, walks the cable at the left after checking the work.

Joseph Rochfort, pictured at left in the yellow safety vest, is an engineer for the New York State Bridge Authority. Rochfort is one of the engineers who give educators and students engineering tours on the bridge. Area educators are stopped at the overlook to learn about the lower structures of the bridge before touring on the bridge.

The museum and classroom space seen here is in the lower level of the administration building at the Bear Mountain Bridge. Area educators, like the New York State Master Teachers here, attend STEM and STEAM workshops to learn how to utilize bridge curriculum in their classrooms. The museum has access to the anchorage and exhibits on bridge engineering and maintenance but is not open to the public.

Engineering intern Juan Cardenas creates videos for STEM classes that excite and educate students about STEM careers. He makes connections between STEM curriculum and STEM careers. He also gives engineering tours of the Bear Mountain Bridge. Seen here with a class of eighth-grade students from Yonkers, Juan explains how suspension bridges work. Being at the bridge instead of just learning in the classroom makes STEM more real for students.

Richy Vacek (front) was the bridge foreman at the Bear Mountain Bridge until his retirement in December 2010. He and his crew stand on the center of the main cable, where it drops below the deck. Vacek, who began at the bridge at the age of 21, created a museum in the bridge administration building. His museum was the inspiration for creating Historic Bridges of the Hudson Valley, whose board he joined upon retirement. With Vacek are, from front to back, Jim Long, Lee Stout, Steve Accilli, Bill Franck, Tom Howley, and Scott Annacone. Maintenance crews on all bridge authority bridges make repairs recommended by the annual inspections of bridge engineers. Vacek's museum, shared with many who visited the Bear Mountain Bridge, is an example of the immense pride felt for their bridges by many at the bridge authority, past and present. (Photograph by John Ellis.)

Four

Kingston-Rhinecliff Bridge

Ferries ran between Kingston, on the west, and Rhinebeck, with a terminal on Rondout Creek, for 200 years, until they were discontinued in 1943 during the war, when gas rationing reduced the number of passengers. Purchasing the *George Clinton*, the bridge authority restored service from 1946 until 1957, when the bridge opened. In 2000, the bridge was dedicated as the George Clinton Kingston-Rhinecliff Bridge. (Hudson River Maritime Museum.)

KINGSTON - RHINECLIFF BRIDGE

The opening of IBM in Kingston and the discontinuation of the ferry generated a major campaign to build a bridge from Kingston to Rhinecliff. In July 1947, the bridge authority appointed consulting engineer David B. Steinman to prepare preliminary surveys and designs for the new bridge. A preeminent suspension bridge engineer, Steinman designed the bridge pictured here. Steinman intended the bridge to cross the Hudson from Kingston Point to just north of the Rhinecliff Station. Bertrand Tallamy, superintendent of the New York State Department of Public Works, was convinced the bridge should be built in a place it could connect the new Thruway and the Taconic State Parkway. The bridge location was moved three and a half miles north. The new location would not support a suspension bridge. The new design would be a continuous underdeck truss bridge.

PROPOSED KINGSTON - RHINECLIFF BRIDGE

The west and east approaches for the new bridge had to be done before foundation work could begin to provide access to the site and accommodate construction equipment. The approaches were staked out by the state engineers, then the trees were cleared, and the right-of-way was rough graded between August 1953 and the summer of 1954.

The east abutment, seen here, would be separated from the west abutment by 30 piers. There were nine land piers on the east and 13 land piers on the west with eight water piers in between. The abutment is the structure at the ends of the bridge, where the bridge structure rests on the ground.

The land piers for the east and west sides of the Kingston-Rhinecliff Bridge are shown here. In the first photograph, the abutment can be seen in the far, top side of the image. The forms for the concrete allow the piers to be poured. To provide for a level grade on the bridge, land piers begin at a point where the land starts to slope to the river. As the piers get closer to the river, they get taller and taller. Men standing near the front piers in the image below gives some perspective to the height of these piers.

Merritt, Chapman, and Scott Corporation, with work on Tappan Zee Bridge completed, floated its batching plant upriver to build foundations for the bridge. The concrete batching plant arrived in September 1954. Water piers of the bridge were built on open caissons. Unlike previous Hudson bridges, the Kingston-Rhinecliff utilized the process of Tremie concrete to fill caissons. Tremie enabled the concrete to be poured into a caisson filled with water. The concrete was poured into a pipe that went to the bottom of the caisson. The bottom cover of the pipe kept the concrete from pouring out until it reached the caisson bottom. Once there, the cover was forced off, and the concrete flowed. As long as the open end of the pipe was below the concrete, it was undiluted by the water. Concrete poured in this way often cured better.

Once the foundation reached the top of the water level, a cofferdam was built around the pier to enable water to be pumped out so the granite block could be laid on the outside of the pier. Rebar seen here was used to reinforce the concrete center of the pier. All substructure work would be complete by November 1955.

Eight water piers crossed this relatively wide section of the river. The underdeck truss would span one main shipping channel on the west side of the main bridge and a minor shipping channel on the east side. Large cranes on floating barges enabled the pier to be built in place. Merritt Chapman, begun as a marine salvage company, would finish here and then go build the foundations for the mighty Mackinac Bridge.

Harris Structural Steel Company began erecting the superstructure when steel arrived on site from Bethlehem Steel in May 1955. The company employed 222 men, including 150 ironworkers, many of whom were Mohawks. Mohawks have worked as ironworkers on many "high steel" projects since the 1880s. They are said to be the best ironworkers on the planet.

Box, or hollow, piers—pier 12 on the west and pier 22 on the east—enable the larger truss structure to connect to the river side of the box pier while the shallow girder section connects to the other side. Box piers contain several interior floors accessed from upper and ground levels. In addition to housing electrical apparatus, they served as an office during construction for workers to pick up paychecks.

The superstructure is almost complete in September 1956; the falsework piers support the erection of the final sections. The completed bridge will have a total length of 7,793 feet; from shore to shore is just under a mile at 5,200 feet. The channel height is 253 feet. Walter Joyce, of Kingston, project engineer for Steinman on the Kingston-Rhinecliff, also worked on the Bear Mountain Bridge and the Roebling bridge over Rondout Creek in Kingston.

Larsen and Johanssen Company, whose company truck is seen here, built the toll booths, administration, maintenance, and garage buildings. The New York State Bridge Authority maintenance on all bridge authority bridges is dedicated to providing the safest bridges and smoothest ride possible. Crews annually clean and paint portions of the bridges and perform minor structural repairs as needed. Five- and 20-year plans provide preventative maintenance well above other transportation authorities.

Gov. Averell Harriman opened the Kingston-Rhinecliff Bridge at 11:30 a.m. on Saturday, February 2, 1957. With some minor construction work still to be completed, the bridge was opened to allow commuters access, since the river was frozen and the ferry would not run. The ribbon was cut by Nancy Heppner, age 13, daughter of bridge authority board member Ernest Heppner. The governor, seen here in ski boots, emphasized the bridge would make it easier and quicker for motorists to make it to the beautiful recreational centers in the mountains. After the governor's speech, a procession of cars crossed the bridge to Route 9G in Dutchess County, where they turned around and returned westward across the bridge. (Both photographs by Howard Gibson, courtesy of Bruce Bohnsack.)

Traffic over the Kingston-Rhinecliff Bridge increased every day following its early opening. With a 36-foot-wide roadway and unlimited clearance for high vehicles, truck traffic was encouraged to use the bridge. The bridge proved a great convenience for the area, particularly to employees of the IBM plant who reside on the east side of the river. (Photograph by Howard Gibson, courtesy of Bruce Bohnsack.)

Dedication of the Kingston-Rhinecliff Bridge was on May 11, 1957. Following a luncheon at the Beekman Arms in Rhinebeck, luncheon guests would form a motorcade and proceed to the plaza on the west end of the bridge for the dedication ceremony. A large crowd heard speeches from Governor Harriman, Judge John Cashin (for Arthur H. Wicks), and Sen. Earnest Hatfield, cosponsors of the original bridge bill in 1947.

Five

Newburgh-Beacon Bridge

On Sunday, November 3, 1963, from 9:00 a.m. to 5:00 p.m., the day after the big dedication ceremonies opening the bridge, the bridge authority ran the Newburgh-Beacon Ferry for one more day to honor the faithful old ferries and their long service. This painting, *The Last Ferry Ride* by artist John F. Gould, portrays the celebration of that day and can be found at www.johngouldart.com. (Robert Gould.)

After more than 30 years of effort to build a bridge to connect the communities of Newburgh and Beacon, the bridge authority contracted with Modjeski and Masters, an engineering firm, to conduct surveys and do preliminary studies and site work, project planning, designs, and cost estimates for the new bridge. Ralph Modjeski, a founding partner, designed the Mid-Hudson Bridge just north of the Newburgh-Beacon Bridge in Poughkeepsie. Gov. Nelson Rockefeller, on this platform on the Hudson River, formally started the drilling machines in a unique "water breaking ceremony" that began the construction of the north span on October 26, 1960. It would take a little more than three years to complete the work on the Newburgh-Beacon Bridge. Concerns were already surfacing about whether or not a two-lane bridge would be able to handle traffic from the developing four-lane highway of Interstate 84.

Substructure work began in October 1960 as a joint venture between Frederick Snare Corporation and the Dravo Company. The three piers in the deepest part of the river were begun first. Caissons used to build the foundations were fabricated in Dravo's plant in Wilmington, Delaware, then towed to Newburgh's South Street Dock before being floated into position and then sunk. Towing distance from the plant was about 270 miles via the Atlantic Ocean. A cofferdam constructed on top of the caisson, seen below, was designed to create a space around the caisson that could be pumped dry. It would allow for piers to be faced with granite block once the concrete had been cured. Work in the river was difficult in cold winter months. Much effort was made to complete the caissons and piers in the warmer months.

Foundations for the river piers, like the one seen here, were originally designed to be open dredged, Tremie sealed caissons. As work progressed, depending on the conditions of the river bottom under each pier, adjustments were made to the process. All 11 river piers were girded with granite masonry from 5 feet below water to 12 feet above to protect the pier from ice damage.

Concrete for all river work was mixed by a floating concrete plant, seen here, owned by Callanan Industries of Kingston, New York. It was equipped with two 56-cubic-foot, end-dump concrete mixers connected to a discharge hopper, which in turn discharged into concrete buckets handled by floating derricks that placed the concrete where it was needed on the pier.

Snare Dravo constructed all truss piers, including two land piers on the east and two land piers on the west, below. Willis and Paul, Inc., was contracted to construct the west approach, which included building three concrete piers, grading, roadway spans, the west abutment, and paving. The east approach, including relocating several houses from the right-of-way, clearing and grading the site, and building three concrete piers and the east abutment, was constructed by the Westcott Construction Company. The land piers were completed by February 1962, and superstructure erection commenced. Steelwork began at pier 1 on the west shore on May 22, 1962. By mid-June, work on pier 15 on the Beacon side was underway, above. Falsework, above, can be seen holding up the truss section during erection as it stretched between piers 14 and 15.

Bethlehem Steel Corporation fabricated 14,920 tons of steel for the bridge superstructure. It was shipped by rail to New Jersey and then by barge to the construction site. Track and work cars on the top of the deck were used to transport steel to the erectors at the end of the span.

Steel was erected by cantilever method from the pier out to a falsework bent. Then the steel erection continued past the falsework, cantilevered out to the next pier. A large floating derrick was used to erect the falsework and the truss panels. Crews worked to erect steel on several piers at the same time.

The bridge over the main channel is a cantilevered through truss, built using two anchor arms on each side. The anchor arms are so called because they hold the ends of the truss that sticks out over the channel. A suspended truss section is erected between the cantilevered sides. This structure allows for a wider span crossing the shipping channel without the need for additional piers to support the structure.

At the toll plaza on the east end of the bridge, a large crane is lifting the structure of the toll booths into place. Taken on December 10, 1962, the photograph shows that winter weather has set in. Winter weather conditions and ice flows on the river will stop construction by the beginning of January.

Concreting on the roadway deck and the approaches resumed in late April 1963. It was necessary to schedule the concreting of the roadway on a specific sequence that would maintain the desired load distribution on the steel trusses. If the contractor concreted the roadway by starting at one end and progressing to the other side, the bridge load would be unbalanced and put too much stress on the concreted trusses.

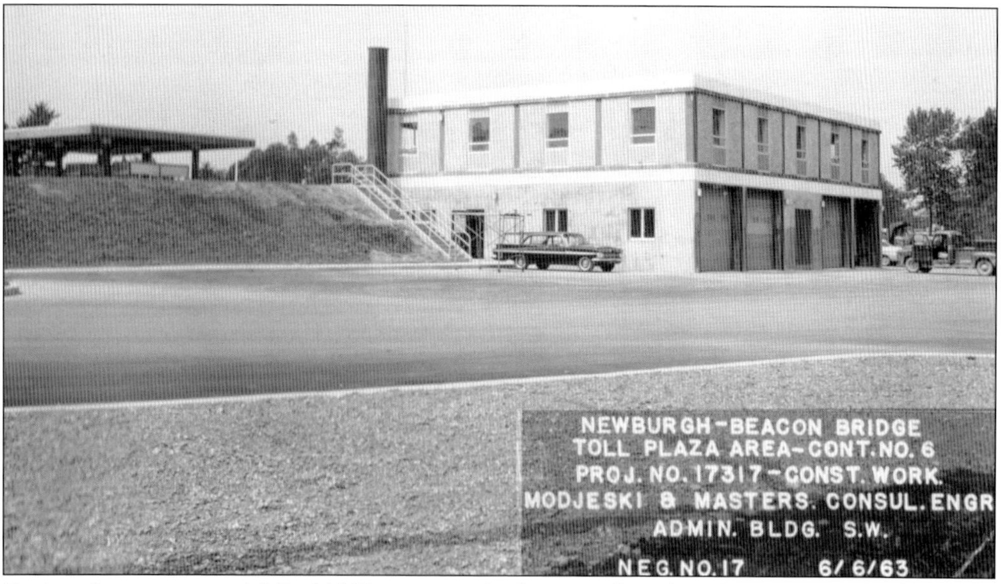

During the summer and fall of 1963, crews worked to complete the administration building and grounds, the toll booths, and approach roads. A utility tunnel had been constructed running from the administration building under the approach to the toll booths and out to the other side. Radiant heating was built into about 200 feet of the approach to the toll plaza to prevent snow and ice hazards.

A massive crane was used to extract the "H" piles supporting the falsework bents once the superstructure was completed. Some of the piles were up to 200 feet long, and the salvage value was very high. Removal of the piles was difficult due to the depth they had been driven. A pile has been pulled from the river, still attached to the crane. Other piles can be seen sticking out of the river along the side of the barge. The piles cannot be left in the riverbed, as they will obstruct the marine traffic. If removed properly, they can be salvaged for future work. Much of the steel truss of the bridge can be seen on the left side of the photograph. The barge carrying the crane is situated under the bridge, and the view looks toward Beacon, on the east side of the river.

The suspended span of the through truss was completed on June 5, 1963. Erection of steel, with the exception of minor members, or pieces, was completed during the first week of August 1963. The final steel erection for the superstructure was on the Beacon approach at pier 7, seen here. Workers celebrated the completion with a flag ceremony.

Looking northwest towards the completed through truss, if one counts the river piers from the far side of the bridge, the traveler on the deck is located above pier 7, where the last truss was just erected. The traveler on the deck and falsework used to hold up that span as it was erected can now be removed from the completed superstructure.

In a speech at the opening of the new bridge on November 2, 1963, linking the cities of Newburgh and Beacon, Governor Rockefeller proclaimed the bridge would spark tremendous economic growth in the region. The Stewart Air Force band played, and Air Force planes flew overhead. The Orange Ferry saluted the bridge from the water. The last ferry ride for the Newburgh Ferry would be the next day. Ribbons were cut at both sides of the bridge. Governor Rockefeller cut the ribbon on the Beacon side at 11:00 a.m., and then the motorcade crossed the new bridge to ceremonies on the Newburgh side. Following the ceremonies, the motorcade traveled to the Newburgh Hotel, where the New York State Bridge Authority hosted a luncheon.

All aspects of the planning, design, and construction for the north span were overseen by New York State Bridge Authority chairman Dr. John Edwards and J. Burch McMorran, chief engineer and later superintendent for the New York State Department of Public Works. Both also worked closely with Ralph C. Holt of Modjeski and Masters, a consulting engineering firm, as well as with the district engineers and staff of all the other contractors responsible for the bridge construction. The resulting bridge was an award-winning design by the Modjeski and Masters engineers. The American Institute of Steel Construction determined the Newburgh-Beacon Bridge to be one of the most beautiful steel bridges of 1963–1964. Dr. Edwards (right) accepted an Award of Merit from Leslie Gillette, representing the American Institute of Steel Construction, on behalf of the entire New York State Bridge Authority.

The first span of the Newburgh-Beacon Bridge, although long-awaited, soon became a frustration. Within the first year after the bridge opened, an average of 25,000 cars crossed it daily. By 1976, vehicle crossing quadrupled. The bridge had become the biggest bottleneck on the river. In October 1975, the Federal Highway Administration gave approval to the bridge authority to begin the design and construction of the parallel Newburgh-Beacon Bridge. By special legislation of Congress, the bridge authority was authorized to collect tolls on the interstate highway bridge for the purposes of paying for maintenance. Above is the Beacon approach with toll booths and traffic, as well as the new span piers in the river. Seen at right is the Newburgh side; the view shows the wider new span, as well as the sidewalk on the south side of the new bridge.

While the bridge is an articulated deck truss structure, the approach spans are held up by girders on piers. The south span's approaches, abutments, and substructure were completed between May 1976 and late 1978. Work was carried out by Gardner M. Bishop; J. Rich Steers, Inc.; Spearin, Preston, and Burrows, Inc.; Yonkers Contracting; and Buckley and Company. Steel-Style, Inc., of Newburgh fabricated the caissons. A caisson at pier 7 tipped when the west side crane broke down. "H" piles were driven into the bottom, and a hydraulic jack righted the caisson in weeks. A truss section in Newburgh is seen in the photograph below.

The steel structure erection was commenced in late 1978 and proceeded through the winter. The bridge's 24,000-ton steel superstructure was built by the American Bridge Division of the US Steel Corporation between early 1979 and mid-1980. Unlike the north span, it was built of Corten steel, a new "weathering" steel, which develops a thin, protective patina of rust and should not need to be painted. Above, maintenance walkways are erected below deck for inspections and maintenance. Below, the west cantilever arm is being erected. Some sections of steel were erected in place, while others were assembled elsewhere and then hoisted into place. A helicopter carried steel members to the bridge to be erected.

The above view is from the west cantilever arm looking outward. The west cantilever arm is part of the through-truss section of the bridge. Below, the west cantilever arm is seen from above. When the west and east cantilever arms are completed, the center suspended section will be set in place. The center suspended section is being assembled in a special assembly yard constructed on the west shore of the Hudson River on the waterfront in New Windsor, just south of the city of Newburgh. Sections assembled in the yard will be floated on a barge to the construction site.

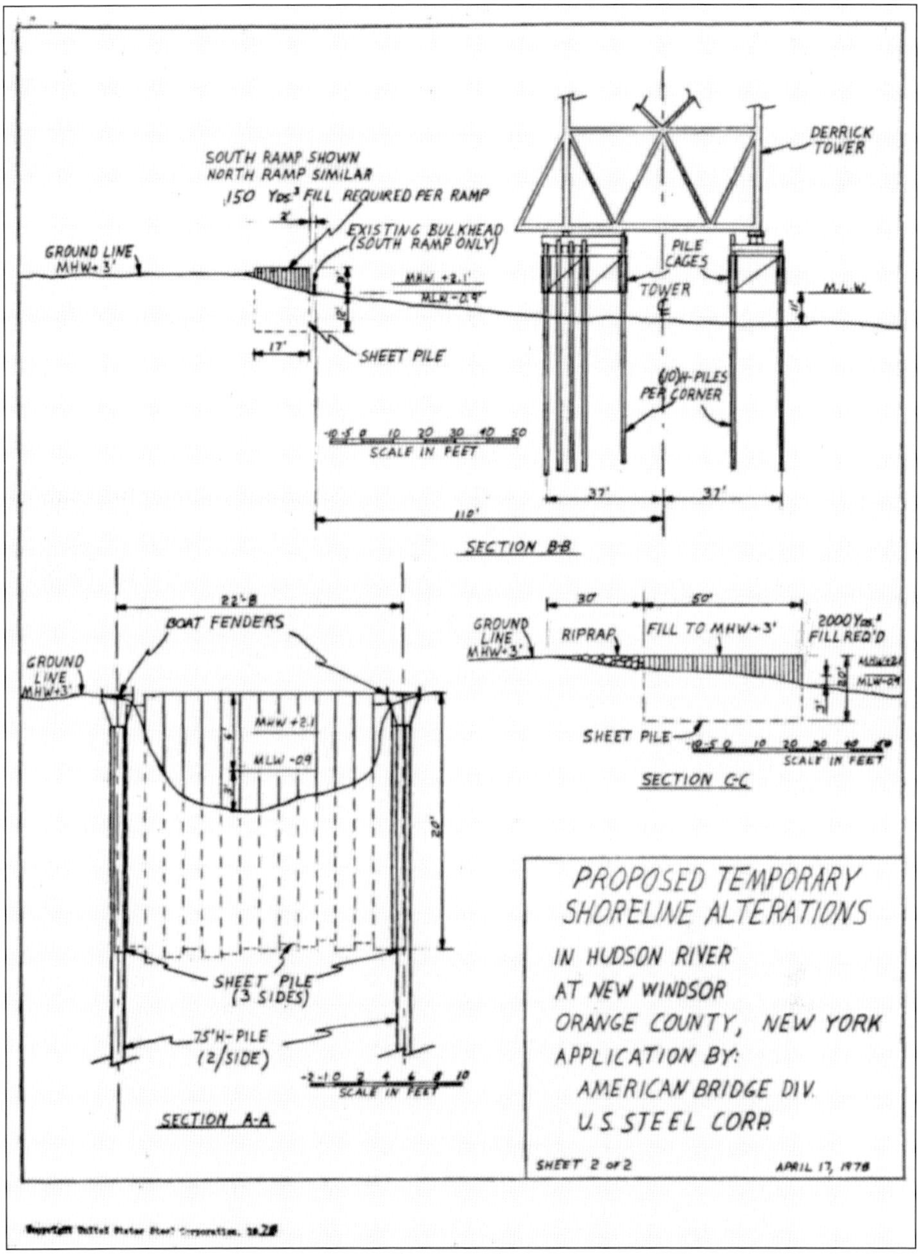

The diagram seen here is part of the application, by American Bridge Company, to build the assembly yard on the waterfront near the bridge construction site. It specifies dimensions and materials that will be needed. American Bridge Company had to prove the yard would be environmentally friendly and not disturb the environment on the shore or in the river. The permit is to build the assembly yard to assemble large sections of the steel structure. Once the work is complete, the assembly yard is to be dismantled. Large sections of the bridge, totaling about 442 feet, were assembled in the yard. A derrick tower and falsework bents were built on barges. The steel sections were built on top of the falsework. Four deck trusses and two anchor spans were floated upriver and installed on the sections of the bridge from the west abutment to pier 4 and from the east abutment to pier 11.

The assembly yard on the New Windsor waterfront is built in an industrial area. The rails for the freight line previously used by Conrail and, later, by CSX run along the river between the assembly yard and River Road. The south span of the Newburgh-Beacon Bridge was the first bridge on the Hudson River to be built by assembling large sections then floating them to the construction site.

Steel is in piles on the shoreline. A derrick with a crane lifts the steel into place, where the pieces are put together like an enormous erector set. Barges in the water bring the steel to where it is needed. The rectangular truss on the bottom is the falsework. An anchor arm is assembled on top. It will be floated upriver at a height almost level with its final position on the bridge.

By March 1980, the south span was 70 percent complete. A "float," a section being moved in place from the assembly plant, occurred in early April and another one in mid-June. The last float, or the last major superstructure lift, occurred on August 20, 1980. A crowd formed on the waterfront to mark this historic event. It would be a little more than two months before the bridge was completed.

The steel structure is in place here in late September 1980. There is still much to be done on the bridge roadway and approaches. Traffic drives by on the north span in the forefront. Forms for the pouring of the concrete, the light-colored area on the right of the south span, are the work of Corbetta Construction, the company that completed the roadway on the Rip Van Winkle Bridge.

A completed expansion dam is shown here above the stringers on the bridge deck. The expansion dam is an expansion joint that allows expansion and contraction of the roadway with variations in temperature. The expansion dam also has means of collecting runoff and channeling it into rigid downspouts, thus protecting the steel structure below from corrosion.

The roadway is ready for the deck pour after the rebar has been installed. A long row of concrete mixers stands behind the low barrier at the far side of the roadway shown here. A long pipe carries the concrete to areas of the roadway. This concrete is a much lighter material than was used on the older bridges. The load on a bridge is minimized by using materials with less weight.

The south span was opened on November 1, 1980, with a motorcade of local officials and dignitaries. An opening road race that began at the Newburgh waterfront and ended at the Beacon riverfront park was the only time pedestrians and bicycles would be allowed on the bridge. The bicycle and pedestrian walkway opened in 1981 for that purpose, the second such crossing on a federal interstate highway.

The completed twin spans stand together here in late November 1980. The lighter-colored silver north span stands out from the dark Corten steel of the south span. When completed in 1980, it was the longest bridge in the world constructed of Corten steel. The hope was that the new bridge would alleviate traffic issues here along Interstate 84 as well as on the other bridges.

Almost immediately after the opening of the south span, the north span was closed to widen it from two lanes to three. About 12,000 tons of concrete and asphalt were removed. Carbon-tipped saws cut the deck into manageable pieces. The asphalt and concrete became part of the fill at the Beacon waterfront. Then the steel superstructure was given a complete overhaul.

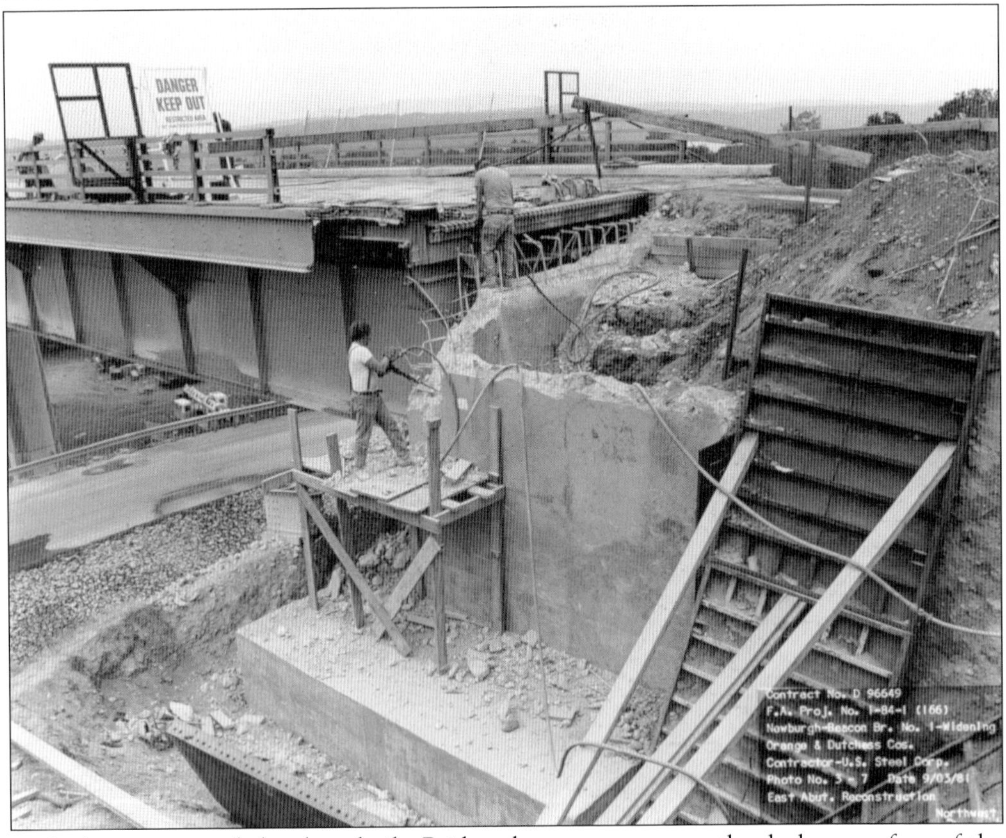

Both abutments needed to be rebuilt. Bridge abutments connect the deck, or surface of the bridge, to the ground and help support its weight both horizontally and vertically. A wider bridge required bigger abutments. The interior section of the abutment can be seen in the photograph, just right of center. It is built like a retaining wall with sides that wrap back around the ground.

Old steel was removed, replaced, and reprocessed. New pins, weighing 500 pounds each, were 13 inches in diameter and replaced those that originally fashioned the critical supports of the old structure, as seen above. The total rehab of the fixed expansion bearings is seen below. Huge jacks support the structure while a new base is inserted below the lifted bearing. The original rivets from the construction of the north span can be seen along the edges of the steel. The old span was completed in 1963, about the time when steel construction was phasing out rivets in favor of welding or high-strength bolting of steel.

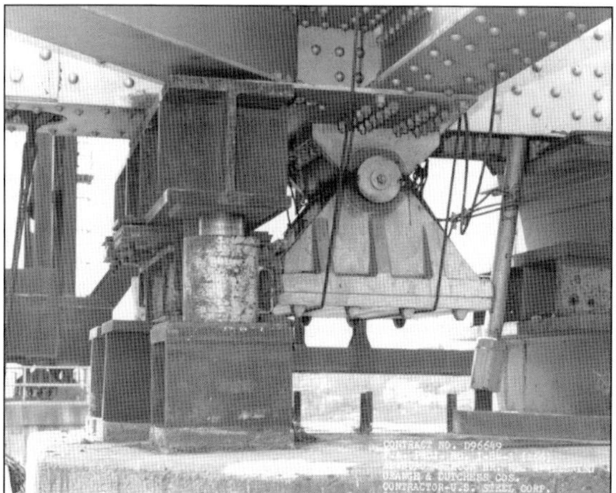

A traveling work platform, seen here, held equipment and materials needed across the structure. The American Bridge Company, a division of US Steel, the company responsible for the erection of the south span, completed the widening of the north span under the direction of Modjeski and Masters engineering firm. Helicopters and barge cranes helped place almost nine million pounds of new steel.

New sub-diagonals were installed to add strength and support to the new wider structure. The darker steel of the new diagonals can be seen in the photograph. The diagonals are traditionally used in construction to strengthen the structure by reinforcing the truss. Cross bracing, inside the truss, adds additional support.

Removing the old steel and adding new and additional stringers will enable the support of the wider structure and extend the life of the bridge. New stringers can be seen here running front to back in the photograph. The older steel stringers from the original construction that are in good shape remain a part of the deck. The bridge deck outside the through truss is widened using stringers cantilevered out from the existing deck.

The south fascia stringer, the stringer at the edge of the deck, is being put in place here. The widening of the deck requires a fascia stringer set farther out on the structure. Within the through-truss section of the bridge, the fascia must be placed right up against the vertical truss.

The new steel installed on the bridge is ready to support a six-and-a-half-inch layer of lightweight concrete brought into position by the concrete conveyor system seen here on 400-foot conveyor belts. The new concrete will be topped with a one-and-a-half-inch dressing of latex concrete, which will provide added protection to the road surface from the pounding it will take from interstate traffic.

The north span was painted to match the south span. To match the dark Corten steel, the north span would be painted brown. The method of painting in this photograph would not be able to be done today. Safety measures on high construction sites are very strict to prevent accidents. But this image shows the massive size of the bridge compared to the workers who gave it a new, longer life.

The north span was reopened on June 2, 1984, after three and a half years of work, with a rededication ceremony on the bridge. Two hundred guests from local, state, and federal governments huddled under the tents for protection from the wind and the rain. The concert band from the Newburgh Free Academy High School played the national anthem, and officials gave speeches.

There was a ribbon-cutting on both sides of the bridge. The crowd and automobiles are gathered here behind the ribbon of the Beacon side of the bridge. Once the ribbon was cut, a motorcade of dignitaries filed across the span, driving down the middle lane of the now three-lane north span of the Newburgh-Beacon Bridge.

The twin spans of the Newburgh-Beacon Bridge stand together, reaching across the Hudson River, one span transporting eastbound traffic, the other span transporting westbound traffic. The north span is 7,855 feet across, and the south span is a bit shorter at 7,789 feet across. It was renamed in honor of Hamilton Fish, who served as New York governor, US senator, and US secretary of state, and subsequent generations of Hamilton Fish who served the Hudson Valley.

The twin spans of the Newburgh-Beacon Bridge were said by New York State transportation commissioner William C. Hennessy to be "not just a bridge, but a whole great transportation complex" at the confluence of Interstates 84 and 87. But more importantly, the Hudson River crossing at Newburgh and Beacon would no longer be the place that truckers using a CB would refer to as "the Interstate 84 parking lot."

Six

The Walkway Over the Hudson

After many long years and several attempts to obtain funds, in 1886, the Poughkeepsie Bridge Company contracted the Union Bridge Company to begin construction of a railroad bridge. Looking northwest from the Poughkeepsie riverfront, the pier construction can be seen in the river. Open caissons were used as foundations for the railroad bridge piers, similar to the Mid-Hudson Bridge about 40 years later. (Walkway Over the Hudson.)

The bridge was built in sections of three cantilever spans connected by two fixed spans that were each 525 feet long. Staging, or scaffolding, was needed to construct the fixed spans. A traveling scaffold used to erect the cantilever spans can be seen on the top left. Two traveling scaffolds moved out from the structure on each side of a fixed span, meeting in the center. (Walkway Over the Hudson.)

In this view looking from the west shore, south of the construction, the railroad tracks run under the new bridge. The use of cantilever construction by means of the traveling scaffold means no interruption of rail service is necessary. The cantilever sections of the bridge were 548 feet long. (Walkway Over the Hudson.)

Sailing ships like the one seen here were used to bring materials upriver to the bridge site. The lumber piled on the shore was used for framing the stagings and the scaffolding, seen in the bridge construction behind the row of trees. Putting up one staging and constructing the steel structure of a fixed span would take three to four months. Steel trusses were transported from Pittsburgh by way of rail lines that would eventually cross the completed bridge. No rail connection would be made with the bridge rails to the north-south lines that run along the river. (Walkway Over the Hudson.)

The nearly completed bridge is seen here in August 1888 from a point north of the bridge on the east shore. Horses and a cart stand next to a pile of lumber and were used to transport heavy materials. Italian immigrant laborers, according to *Poughkeepsie News-Telegraph*, the newspaper at the time, were not taking jobs away from Americans, as was sometimes claimed. They often did the hard and menial work that no one else could or would do. These laborers received $1 to $1.25 per day. (Walkway Over the Hudson.)

While railroads were the transportation means of choice on the land, steamboats were the favorite way to travel on the river. Day liners and sleepers can be seen here carrying people to destinations up and down the Hudson River. The Hudson River ferry boat *Brinckerhoff* can be seen in the water on the left. The ferry would transport people across the river, while the other boats would travel north and south. The city of Poughkeepsie can be seen on the shore on the right. (Walkway Over the Hudson.)

When the Poughkeepsie Bridge opened in 1889, it was the longest bridge in North America and the first bridge to cross the Hudson River between New York City and Albany. The bridge became a key rail hub linking the eastern industrial centers with the raw materials to the west. Both freight and passenger trains, until 1955, used the Poughkeepsie Bridge. By 1905, over 400 freight cars crossed the bridge each day. By the early 1920s, nearly 2,000 cars crossed each day. Trolley cars were modified to run on both trolley tracks and railroad tracks and transported people from New Paltz to Poughkeepsie. During the wars, there were armed patrols protecting the bridge. (Walkway Over the Hudson.)

The Poughkeepsie Bridge connected rail service throughout the Northeast for over 85 years before service ended. The last train, pictured above, rolled across the tracks above the Hudson River on May 8, 1974. It was headed eastbound from Maybrook, New York, en route to New Haven, Connecticut. The last train on the bridge can be seen here from Waryas Park on the Poughkeepsie shoreline. In the photograph to the right, the old railroad ties look dry and covered with residue from years of train traffic. Due to cuts in funds, there was no longer any railroad trackwalker paid to check the tracks for sparks once a train has passed over the bridge. (Both, Walkway Over the Hudson.)

At about 2:45 p.m. on May 8, 1974, hours after the last Penn Central train had crossed the bridge, smoke and flames could be seen breaking out on the bridge. From the distance of the Marist College campus in Poughkeepsie, the fire looked as it does in the photograph above. Firefighters had no equipment that could reach the bridge deck. Firefighters in the distance can be seen in the photograph below, helpless to put out the fire. Dry weather conditions and the creosoted timbers made the fire worse. The heat was so intense it damaged and warped the steel track supports, and pieces fell from the bridge. (Both, Walkway Over the Hudson.)

The abandoned Poughkeepsie Bridge can be seen here. A single track is in the center of the bridge where, for the first almost 30 years, there were two tracks. The second set of tracks was removed to make room for a gauntlet track, also known as an interleaved track. It was installed to handle the weight of diesel locomotives but was removed in 1958. (Walkway Over the Hudson.)

With the debris removed, the steel infrastructure remains sound after almost 120 years. Seeing the bridge here, it is easy to see why the Walkway Over the Hudson group was founded. Unobstructed views up and down the Hudson River are possible from the bridge's location 200 feet above the water. To enlist community support for the project, access was built to provide tours from the Highland side of the bridge. (Walkway Over the Hudson.)

After many years and changes to the Walkway Over the Hudson group, and as of January 2004 led by Fred Schaeffer, the group was able to make headway and begin to get some serious funding. Donations from the Dyson Foundation funded a significant portion of the design and inspection. Clean-up was done of the old railroad rails, and timbers and railings of the old bridge had to be removed. The tracks were the last to go, as they helped haul the other materials off the bridge structure. In the photograph below, the old electric lines that hung off the bridge have been removed. The steel structure that will support the new walkway remains. (Both, Walkway Over the Hudson.)

The bridge clean-up seen from below gives a whole different perspective of the work. More than 200 feet above the water, with no barrier on the sides, workers needed to be secured to the structure. The removal of the old material from the bridge was done by Environmental Remediation Services. (Walkway Over the Hudson.)

Engineering inspectors rappel down the pier from the deck to inspect the steel of the piers. Most of the steel is still good, but additional trusses and repairs will be needed to support the weight of the deck of the walkway and ensure the bridge is safe in strong winds. (Walkway Over the Hudson.)

Workers climb around the steel structure of the old bridge, carefully and systematically bringing it back to life as the walkway. It is estimated that nearly one million rivets were used to hold the 20,000 tons of steel together. Each of the four piers holding up the bridge contains 40,000 tons of timber, stone, and concrete. Most of the concrete used on the original bridge is Rosendale concrete, from nearby Rosendale to the west of the bridge. It is hard to imagine not taking a great deal of time enjoying the view if one were the worker in the photograph below. The Mid-Hudson Bridge, seen to the south, turns 90 in 2020 but is a much younger structure than the old railroad bridge. (Both, Walkway Over the Hudson.)

A ground-breaking ceremony was held in May 2008 as construction work began to transform the old railroad bridge into a park. Seen here holding souvenir spikes are, from left to right, Congressman Maurice Hinchey; Ned Sullivan, president of Scenic Hudson; Rob Dyson, chairman of the Dyson Foundation; Carol Ash, board member of Open Space Institute and the Mohawk Hudson Land Conservancy; James Sproat, director of New York State Office of General Services; Judith Enck, deputy secretary for the environment in the New York Governor's Office; Sen. Stephen Saland; and Fred Schaeffer, attorney at law. Schaeffer, seen below with his ever-present camera, has been a major driving force to make the Walkway Over the Hudson a reality. (Both, Walkway Over the Hudson.)

The west side of the bridge travels through a wooded area with a soft curve to the tracks. Entering the walkway from the Highland side, the trees offer shade and protection from the wind. It will be a wonderful place for a leisurely stroll. (Walkway Over the Hudson.)

The east side of the bridge rises over the city of Poughkeepsie, giving a different kind of a view with more activity and buildings, old and new. Below the bridge is the beautiful campus of Marist College. Near the waterfront is the Mid-Hudson Children's Museum. The walkway will draw people from the region and all over to visit the area. (Walkway Over the Hudson.)

Ironworkers continue to make improvements to strengthen and reinforce the steel structure. Three overlook areas must be added to the bridge where the walkway will expand to be three and a half feet wider on each side to allow for scenic viewing in an area out of the main flow of visitors. The finished deck will be 25 feet wide and will have a four-foot, six-inch railing. Crews are working on many parts of the bridge at once, and the work is moving along on schedule. (Both, Walkway Over the Hudson.)

Resembling an enormous erector set, the steel pier (left) stands gracefully high above the Hudson. Painted black during World War II to make it less visible in the event of an attack, the bridge continued to be painted until the 1960s. Metal reconstruction experts claim the absence of paint actually helped the steel stay in the good condition it is in today. The scaffolding hung around the bottom of the pier, as seen at left, gives the impression in the photograph below that the workers are floating on air above the Hudson River. There is undoubtedly no better place to have lunch than a spot with the incredible view of the river, the valley, and the iconic Mid-Hudson Bridge. (Both, Walkway Over the Hudson.)

Engineering and design for the Walkway Over the Hudson was overseen by Bergmann Associates of Rochester, New York. New panels were fabricated by the Fort Miller Company, Inc., from Schuylerville, New York. Installation of the concrete panels began in September 2008 and was completed within one year. The walkway surface was constructed using 973 prefabricated concrete panels. It started at the east and west sides of the bridge to balance the weight as the panels were laid in place. One by one, work progressed until the two sides met in the center of the bridge. The last panel of the walkway is being lifted into place below. Prior to being permanently set, it was signed by all who helped to make the Walkway Over the Hudson a reality. (Both, Walkway Over the Hudson.)

The opening of the Walkway Over the Hudson was a special part of the Hudson-Fulton-Champlain quadricentennial in 2009. The replica of Henry Hudson's ship the *Half Moon* sailed under the walkway as part of the year-long festivities organized by quadricentennial executive director Tara Sullivan. (Walkway Over the Hudson.)

Opening the beautiful Scenic Hudson gates ceremonially opened the Walkway Over the Hudson on October 3, 2009. Scenic Hudson was instrumental in transforming an old railroad bridge into an incredible elevated park. The reconstruction project cost $38.8 million and was funded by small and large donations and grants. Once the project was complete, New York State Parks assumed responsibility for public use of the 6,768-foot walkway. (Walkway Over the Hudson.)

In the first 18 months after its opening, 850,000 people visited the Walkway Over the Hudson. There is no fee to enter the park, which is open and staffed from 7:00 a.m. to sunset. Visitors walk, run, bike, and walk dogs on a leash on the walkway. Many local residents visit the walkway on a regular basis. Of the three million visitors to the park in the first 10 years since it opened, almost half surveyed come from outside Dutchess and Ulster Counties, which border the park. Visitors have come from all 50 states and 42 countries. Visitors to the park often speak with one of the park's many ambassadors. The program, begun in 2010, was initially funded by an award from the Cunneen Hackett and McCann Foundations. Ambassadors are volunteers, and many were part of the original walkway group. (Both, Walkway Over the Hudson.)

In 2010, the walkway ownership transferred to the New York State Bridge Authority. The bridge authority owns and maintains the bridge structure and the piers. Bridge authority maintenance is seen here checking the connection points and expansion joints between sections on the bridge. In the bucket are, from left to right, then–chief engineer Bill Moreau, Ryan Griffiths, and Rick Warr. On the truck is Jared Guess, and on the bridge is Craig Gardner.

Katherine Kelly, professional engineer and underwater diving inspector, is at a pier of the walkway. Federal guidelines require bridge piers to receive hands-on underwater inspection on a five-year cycle to identify defects like rotted or split timber, marine borers, undermining of piers, rusted steel elements, cracking, and more. Challenging conditions of the Hudson River include limited- to no-visibility water, fast tidal flows, local marine life, and potential entanglement. (W.J. Castle PE & Associates.)

The iconic Mid-Hudson Bridge and the incredible Walkway Over the Hudson stretch across the waters of the majestic Hudson River. Both bridges were built to connect. The Poughkeepsie Railroad Bridge was built to connect materials in the west with factories in the Northeast. The Mid-Hudson was built to connect people and places on the west side of the river with people and places on the east side. Since the making of the walkway on the old railroad bridge, the connection is no longer limited to crossing the river. The two bridges have been made into a walkway loop, about four miles of an active and beautiful walking trail. Only in the Hudson Valley would one find such a unique variety of bridges that are so well maintained for the lowest tolls of any bridges in the country. (Walkway Over the Hudson.)

Consistent with our mission to preserve history on a local level, this book was printed in South Carolina on American-made paper and manufactured entirely in the United States. Products carrying the accredited Forest Stewardship Council (FSC) label are printed on 100 percent FSC-certified paper.

MADE IN THE USA